Software Design Descriptions

A How To Guide for Project Staff

David Tuffley

To my beloved Nation of Four
Concordia Domi – Foris Pax

Design is a plan for arranging elements in the best way to achieve a purpose. David Tuffley

Acknowledgements

I am indebted to the Institute of Electrical and Electronics Engineers on whose work I base this book, specifically IEEE Std 1016 Recommended Practice for Software Design Descriptions.

I also acknowledge the *Turrbal* and *Jagera* indigenous peoples, on whose ancestral land I write this book.

Contents

Contents

A. Software Design Descriptions

A.1. Scope

This standard provides a guide for writing Software Design Descriptions (SDDs). It describes:

- the software development context in which an SDD should be created

- minimum requirements for SDD format and content and

- the qualities of a good SDD.

This standard represents a tailoring of the *IEEE Std 1016 Recommended Practice for Software Design Descriptions*. SDDs prepared in compliance with this standard will therefore also comply with IEEE Std 1016.

B. Definitions, acronyms & abbreviations

Design Entity: an element (component) of a design that is structurally and functionally distinct from other elements and that is separately named and referenced (IEEE Std 1016). An entity should describe a discrete system component that can be considered, implemented, changed and tested with minimal effect on other entities. Example: Software Module Detailed Design Description, Database Design, Memory Common Design, Interface Design.

Module: a program unit that is discrete and identifiable with respect to compiling, combining with other units, and loading; for example the input to or output from, an assembler, compiler, linkage editor, or executive routine (IEEE Std 610.12 - Standard Glossary of Software Engineering terminology). Note that the terms module, component and unit are used interchangeably.

Program Step: a single executable instruction in the target language. Program statements, selections and iterations are classed as single program steps. For example:

```
If {expression} then       Step 1       (selection)
        {instruction}      Step 2       (statement)
        else               Step 3       (selection)
        {instruction}      Step 4       (statement)
end if
Do while {expression}            Step 1
    (iteration)
            {instruction}              Step 2
    (statement)
end do
```

Note that compound statements such as:

```
{instruction} ; {instruction} ; {instruction}
```

are classed as multiple program steps.

SAS: System Architecture Specification - provides a framework for the development of the detailed design. It describes the partitioning of the design solution into individual hardware and software modules and describes the interaction between these modules. It does not provide detailed descriptions of the working of each module. This is the function of the Software Design Description.

SRS: Software Requirements Specification - a statement of what the system must do from the user's point of view.

SDD: Software Design Description - a detailed description of the workings of each software component.

SDL: State Definition Language.

Structure Chart: a diagram that identifies modules, activities or other entities in a system or computer program and shows how larger or more general entities break down into smaller, more specific entities.

System Architecture: a collection of hardware and software components and their interfaces that constitute the system.

White Box Testing: testing that takes into account the internal structure of a body of software thus ensuring that all branches and paths are traversed and all statements executed.

C. Purpose

The SDD is created by the System Architect or designer and is the major deliverable from the detailed design process.

C.1. Prerequisites

The prerequisite document required for an SDD varies according to the size and complexity of the software product to be developed (refer figure 1).

Large Systems. In the case of large and complex systems the prerequisite for the development of an SDD is the System Architecture Specification. In this context the SDD represents a further refinement of the design entities described in the SAS. An SDD may provide descriptions of one or more design entities.

Small Systems. In the case of small systems the SDD prerequisite is a Software Requirements Specification. In this context it is the single source of design solutions to problems stated in the SRS.

C.2. Use

The SDD is the primary reference for code development. As such, it must contain all the information required by a programmer to write code.

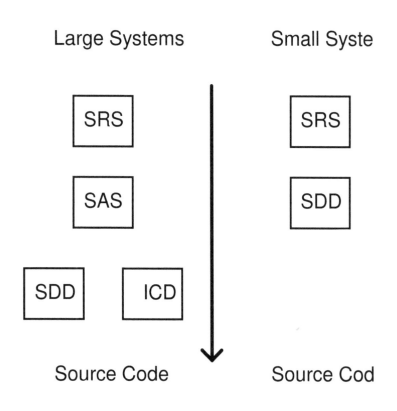

D.Quality criteria

This section addresses the criteria by which the quality of an SDD may be judged.

D.1. Traceability

Each design entity described in the SDD must satisfy a specific requirement specified in the Software Requirements Specification or an internal design requirement (i.e. an internal need for portability of a software product across many operating systems may create the need for an operating system services interface module).

D.2. Standards compliance

The SDD must address each issue specified in this standard. In practical terms this means that:

- all SDDs shall have the paragraph headings specified below

- if a paragraph heading is not applicable to the design it is to be included and marked not applicable

- all applicable paragraphs shall address the checklist of key points covered in each subsection of this standard.

D.3. Completeness

- Are all the requirements satisfied by design entities?
- Are all module input parameters used somewhere in the module?
- Are all output parameters assigned a value somewhere in the module?
- Does the design address:

    - Power fail/restart

    - Security

    - Reliability

    - Maintainability

    - Dealing with and recovering from hardware failure (i.e. disk, communications link).

D.4. Comprehensibility

- Are design concepts presented in simple, unambiguous language?

- Can design concepts be understood on the first reading?

- Does the design description state the obvious, leaving nothing to the imagination of the reader?

D.5. Modularity

This is the degree to which the big problem is broken up into smaller sub-problems, whose solutions may be pursued independently.

- Does each module deal with one simple self contained problem?

- Can a reviewer understand the module design on the first reading with only limited understanding of the complete system?

- Will the module design translate to a maximum of 100 program steps?

- Can the module be tested in isolation?

D.6. Cohesion

This is the degree to which internal elements of a module are related to each other.

- Can the module's function be described in one sentence with one subject and one verb?

- Does a module perform only one or two related functions.

Example:

Poor Cohesion - The Control Module detects incoming messages and passes them on to the appropriate message server, runs any time initiated tasks due, checks for failed communications links and attempts to resume communications.

Good Cohesion - The Control Module runs tasks a predefined times of the day.

D.7. Coupling

This is the degree to which individual modules are independent from each other.

Good - Data coupling. (A maximum of 5 parameters are passed between modules.)

Less Good - Control coupling. (One module controls the sequencing of steps in another module.)

Bad - Common coupling. (Several routines reference a single common data block.)

Very bad - Content coupling. (One module modifies local data values in another module.)

D.8. Fan-in

This is the number of superordinate modules that call a module.

- Maximise fan-in for good modularity
- Low fan in - absorb into other modules.

Fan-out (span of control)

This is the number of subordinate modules called by a module. A high fan-out (>7) may indicate that a module is too complex. That is, it contains too much control and coordination logic.

D.9. Implementability

The SDD must provide all information that the programmer requires to code the entity described.

D.10.Modifiability

Design information will be presented in one place only to support future modification of the document. An instance of more than one occurrence of the same information may result in ambiguity if all occurrences of the information are not updated simultaneously.

D.11.Extendability

This is the ease with which the design may be adapted to changes in requirements. Key elements of extendability are:

- Design simplicity

- Modularity. The more autonomous the modules in a software architecture, the higher the likelihood that a simple change will affect just one module rather than trigger a chain reaction across the whole system.

E. Format & content

This section provides the minimum requirements for format and content of the SDD. All sections described here must be provided in SDD documents. Issues relevant to particular designs that are not provided by this outline may be included as required by the Designer and approved by the Project Manager. Figure 2 provides the standard SDD Document Outline.

E.1. Cover page

The cover page is to include:

- Project Title
- Document Number
- Volume Number
- Document Class: `Software Design Description'
- System Name
- Revision Number
- Author
- Date printed

- Approval Authority
- Approval Date

E.2. Page marking

Each page is to be marked with:

- Document Class: `Software Design Description'
- System Name
- Document Number
- Revision Number
- Page Number

```
1. Introduction
     1.1 Purpose
     1.2 Scope
     1.3 Definitions, acronyms and abbreviations
2. Reference
3. Design decomposition
     3.1 Design structure
     3.2 Code structure
4. Module detailed design description
     4.n Identification
          4.n.1 Purpose
          4.n.2 Function
          4.n.3 Subordinates
          4.n.4 Dependencies
          4.n.5 Interface
               4.n.5.1 Input Assertion
               4.n.5.2 Output Assertion
               4.n.5.3 Interruption
               4.n.5.4 Uses
               4.n.5.5 Used By
```

```
                    4.n.5.6 Calling Sequence
                    4.n.5.7 External Data
            4.n.6 Resources
            4.n.7 Internal Data
            4.n.8 Processing
                    4.n.8.n Sub-module Name
                            4.n.8.n.1 Function
                            4.n.8.n.2 Interface
                            4.n.8.n.3 Resources
                            4.n.8.n.4 Internal Data
                            4.n.8.n.5 Processing
                            4.n.8.n.6 Implementation
                                        Process Info
            4.n.9 Implementation Process Info
            4.n.10 Unit Test strategy

5. Interface design description
        5.n Identification
        5.n.1 Type
        5.n.2 Purpose
        5.n.3 Function
        5.n.4 Protocol
        5.n.5 Errors
        5.n.6 user interaction

6. Database design
        6.1 Introduction
        6.2 Data Dictionary
        6.3Database design

7. Global memory

8. Messages
        8.1 Introduction
        8.2 User messages
        8.3 Interprocess messages

9. Index
```

Figure 2. Software Design Description Document Outline.

17

E.3. Introduction (section 1)

The purpose of this section is to provide an overview of the purpose and scope of the system, the design of which is described in this document.

E.3.1. Purpose (section 1.1)

- Identify the system name
- Describe the reason for the creation of the system
- Identify the intended audience for the document.

E.3.2. Scope (section 1.2)

- Describe the functions satisfied by the SDD. Be brief - provide one sentence per function.
- Refer to the specification satisfied by the SDD (example: Software Requirements Specification or System Architecture Specification).
- Identify any functions that are outside the scope of the design (i.e. external protocols, operating system services).

E.3.3. Definitions, acronyms & abbreviations (section 1.3)

Describe terms used that cannot be expected to be common knowledge within the document's target audience.

Exclude data entity and attribute names that are to be included in data dictionaries.

E.4. References (section 2)

Identify the preceding documents from which this document was created; for example Software Requirements Specification XYZ.

Identify all documents referenced in this document specifying reference code (to be used in the body of the text), document name, author, version number, release date and location.

Example:

[1] XYZ Software Requirements Specification V1.04, 23-9-1991, J. Bloggs, Data Bank

E.5. Design decomposition (section 3)

This section provides a road map to the physical code in terms of the code structure and the set of physical files that will contain all source code.

If the SDD does not provide a further decomposition of design described in the SAS it is noted here and reference made to the SAS. For example, this will occur where the SDD describes the implementation of a single design entity introduced in the SAS.

E.5.1. Design structure (section 3.1)

Provide an overview of the design solution in terms of a set of named design entities and their associated interfaces.

Classes of design entities are:

- Software Module Detailed Design Description

- Database Design

- Global Memory Description

- Interface Design.

Provide a diagram depicting the design decomposition.

Examples:

- Finite state diagram (i.e. SDL diagram)

- Data flow diagram

- Structure chart.

E.5.2. Code structure (section 3.2)

Provide a:

- list of files that will contain all code described in the design.

- list of include files.

- list of standard library files.

E.6. Module detailed design description (section 4)

The SDD shall provide a module detailed design description for each physical software module described in the design decomposition. Each module description shall include data described in the following sections.

E.6.1. Identification (section 4.n)

Identify the software module name in the section heading.

Example: 4.1 Message Routing Module

E.6.1.1. Purpose (section 4.n.1)

The reason for the module's existence including the user or design requirement that is satisfied (refer to the paragraph number of the related SRS or SAS).

E.6.1.2. Function (section 4.n.2)

A summary statement of what the module does.

E.6.1.3. Subordinates (section 4.n.3)

Identify all the design entities that comprise this entity (for example, function descriptions, subroutine descriptions and data structures local to this entity). This information is used to document parent/child structural relationships through a software system decomposition.

These relationships may be described by reference to a structure chart.

E.6.1.4. Dependencies (section 4.n.4)

This is a description of the relationships of this module with other modules.

Example:

Identification of other design entities, that must exist for this module to function (i.e. Co-operating processes, shared memory)

E.6.1.5. Interface (section 4.n.5)

This is descriptions of the nature of interactions between entities that comprise this module.

E.6.1.5.1.   Input Assertion (Section 4.n.5.1)

These are assumptions made by the module that must be true for the module to execute correctly. Assumptions may include consideration of timing, Sequencing/Order of Execution.

Example: The module assumes that the network node address passed as a parameter from the calling module is a valid address.

E.6.1.5.2.Output Assertion (Section 4.n.5.2)

This is a statement of what the module outputs will be.

Example: The module outputs a 200 byte message string consisting of network node address (4 bytes) + alpha/numeric message text (196 bytes).

E.6.1.5.3.Interruption (Section 4.n.5.3)

Describe the process by which this module can be interrupted (i.e. hardware, other tasks) and provide the interrupting module names.

E.6.1.5.4.Uses (Section 4.n.5.4)

Specify the names of all modules that are called by this module. These relationships may be described by reference to a structure chart.

E.6.1.5.5.Used By (Section 4.n.5.5)

Specify the names of all modules that initiate/call this module. These relationships may be described by reference to a structure chart.

E.6.1.5.6.Calling Sequence (Section 4.n.5.6)

Arguments. Describe the arguments passed to the module and those returned from the module. For each argument provide:

- name
- description
- purpose - data import, export or both
- data type
- valid range of values

Command Sequence. If an argument passes a command, describe valid command sequences (i.e. open-que, en-que)

Message Passing. Format and content of messages passed.

E.6.1.5.7.External Data (Section 4.n.5.7)

References to data external to this module (i.e. environment variables, shared memory, global data references).

Note: All parameters and message components must be explicitly described in the Data Dictionary.

E.6.1.6. Resources (section 4.n.6)

A description of the elements used by the entity that are external to the design; for example:

- printers, plotters, chart recorders
- disk partitions
- memory allocation, buffers
- CPU cycles
- operating system services (i.e. task to task communication facilities)
- program library services (i.e. maths functions).

E.6.1.7. Internal Data (section 4.n.7)

For all data internal to the module provide:

- name
- description/use
- data type

- initial value
- valid values.

E.6.1.8. Processing (section 4.n.8)

A description of the rules to be used by the module to achieve its function:

- algorithm descriptions
- processing steps
- calculations
- process termination conditions
- management of exception conditions
- interrupt handling.

Processing may be expressed in terms of structured English/pseudo code (refer to Appendix A), flow charts, Nassi-Shneiderman diagrams, state transition diagrams.

If the module consists of submodules describe each submodule in terms of the following subsections:

E.6.1.8.1.Sub-module Name (Section 4.n.8.n)

Function (Section 4.n.8.n.1) - Refer Section 5.6.3.

Interface (Section 4.n.8.n.2) - Refer Section 5.6.6.

Resources (Section 4.n.8.n.3) - Refer Section 5.6.7.

Internal Data (Section 4.n.8.n.4) - Refer Section 5.6.8.

Processing (Section 4.n.8.n.5) - Refer Section 5.6.9.

Implementation Process Information (Section 4.n.8.n.6) - Refer Section 5.6.10.

E.6.1.9. Implementation process information (section 4.n.9)

Provide information required by the programmer to code this module.

Example:

- Identification of code templates to be used.

- Identification of an existing module from which this module can be derived.

- Identification of standard include files.

- References to coding standards.

- Specification of coding practice.

- The name of the file that is to contain the code.

E.6.1.10. Unit test strategy (section 4.n.10)

Provide a strategy for the effective white box testing of the module. The strategy should provide the following test data:

- **Test environment**. Describe the test software required to perform the test (i.e. module drivers and stubs).

27

- **Input Data.** Provide an input data set.

- **Output Data.** Describe expected output data.

- **Test Procedure.** Describe the set of steps required to set up, start, proceed with, suspend and stop the test.

- **Pass/fail.** Describe test pass/fail criteria.

E.7. Interface design description (section 5)

The SDD shall provide an interface design description for each interface described in the design decomposition. Interface designs describe the interactions between:

- system modules

- system modules and external devices (i.e. programmable logic controller)

- human beings and system modules.

- Interface descriptions shall include:

E.7.1. Identification (section 5.n)

Identify the interface name in the section heading.

Example: 5.2 Workstation User Interface

E.7.1.1. Type (section 5.n.1)

Describe the type of interface; for example - task to task, external, user interface.

E.7.1.2. Purpose (section 5.n.2)

This is the reason for the interface's existence including the user or design requirement that is satisfied (refer to the paragraph number of the related SRS or SAS).

E.7.1.3. Function (section 5.n.3)

A description of what the interface does.

E.7.1.4. Protocol (section 5.n.4)

A description of the communications protocol incorporating:

- Message format and description
- Message component names (refer Data Dictionary for data descriptions, types and expected values)
- Message initiation
- Fragmentation and reassembly of messages
- Error detection, control and recovery procedures
- Synchronisation, including connection establishment, maintenance, termination and timing

- Flow control, including sequence numbering, window size and buffer allocation

- Data transfer rate

- Transmission services including priority and grade

- Security including encryption, user authentication and auditing

E.7.1.5. Error codes (section 5.n.5)

As required.

E.7.1.6. User interaction (section 5.n.6)

For all user interfaces include:

- User inputs

- Screen formats

- Interaction dialogue

- References to error and information messages.

Note that, at the designer's discretion, complex interfaces may be described in overview only. Detailed descriptions may then be provided in succeeding Interface Control Documents (ICDs).

E.8. Database design (section 6)

This section provides descriptions of all databases that are global to two or more software modules. Structures described here are referenced in all software module design descriptions to avoid redundant presentation of design information.

E.8.1. Introduction (section 6.1)

This section introduces the types of data structures that are to be employed in the design (i.e. relational, hierarchal).

E.8.2. Data Dictionary (section 6.2)

This section provides a definitive reference of all data entities that are global to two or more software modules. Entities are described in terms of:

- name
- description
- data type
- format
- valid values.

E.8.3. Database design (section 6.3)

This section provides a data base design using a design methodology appropriate to the target data base facility. As a minimum requirement databases are described in terms of:

- entities
- entity attributes
- relationships between entities
- modules that access the entity
- modules that change the entity.

E.9. Global memory (section 7)

This section provides the format and content of global memory. Global memory elements shall be described in terms of:

- name
- description
- data type
- format
- valid values
- modules that access the element
- modules that change the element.

E.10.Messages (section 8)

This section provides descriptions of the format and content of all messages that are passed either internally between modules by task to task communication facilities or externally to users by user interface facilities. Message formats described here are referenced in all software module design descriptions to avoid redundant presentation of design information.

E.10.1. Introduction (section 8.1)

This section introduces the concept of a single reference for all system messages.

E.10.2. User messages (section 8.2)

This section describes all user information messages (including error messages) in terms of error message code and message text. Note that it is standard design practice that all messages shall be stored as data in files or database structures. Message text shall not be embedded in code.

E.10.3. Interprocess messages (section 8.3)

This section describes the format and content of all interprocess messages.

E.11.Index (section 10)

This section shall provide a quick reference to key issues addressed by the SDD. A minimum of two levels of indexing shall be provided.

Example:

Commands

open, 14, 140-145
close, 23, 191-195
put, 35

F.   Appendix A - Structured English

All processing can be described in terms of three basic processes: sequence, decision and iteration.

Sequence

A pure sequence process occurs when one action is followed by another. Sequence is described in structured English as successive lines of text, each line representing one action:

```
action 1
action 2
```

Example:

```
retrieve hourly rate
retrieve hours worked in current week
compute actual spent this week
```

Decision

Decision processes are described in terms of if-then, if-then-else, or case statements:

```
if-then

if condition then
    action
end-if

Example:

If meter count greater than 0 then
compute total charge
end-if

if-then-else

if condition then
    action 1
else
    action 2
end-if
```

Example:

```
If meter count equal to or greater than 0 then
    compute total charge
else
    send error message
end-if

case

case selector of

    condition 1: action 1
    condition 2: action 2
    condition 3: action 3
    condition n: action n
end case
```

Example:

```
case customer class of
    low volume: selling price = list price
    medium volume:selling price = list price -10%
    high volume:selling price = list price - 20%
end case
```

Iteration

There are two classes of iteration – while-do and repeat-until.

while-do

In the while-do construct the first line contains the condition to be tested at the beginning of each iteration of the actions within the block:

```
while condition do
    action
end-while
```

Example:

```
while message buffer empty flag is clear do
    process message
    purge message from buffer
    if no more messages then
        set message buffer empty flag
```

```
        end-if
end-while
```

repeat-until

In the repeat-until construct, the keyword repeat
appears at the beginning of the repeat-until block
to mark the start of the block. The actions are
listed and indented below the repeat line. The last
line in the block contains the condition to be
tested to determine if the actions within the
iteration block should be repeated:

```
repeat
    action
    until condition
```

Example:

```
repeat
    read user password
    if user password is not valid then
        print invalid password message
    add one to number of tries
    end-if
until user password is valid or number of tries = 5
```